NATIONAL DEFENSE RESEARCH INSTITUTE

# Moving to the Unclassified

## How the Intelligence Community Can Work from Unclassified Facilities

Cortney Weinbaum, Arthur Chan, Karlyn D. Stanley, Abby Schendt

Prepared for the National Geospatial-Intelligence Agency

For more information on this publication, visit www.rand.org/t/RR2024

**Library of Congress Cataloging-in-Publication Data** is available for this publication.
ISBN: 978-1-9774-0048-2

Published by the RAND Corporation, Santa Monica, Calif.


The RAND Corporation is a research organization that develops solutions to public policy challenges to help make communities throughout the world safer and more secure, healthier and more prosperous. RAND is nonprofit, nonpartisan, and committed to the public interest.

RAND's publications do not necessarily reflect the opinions of its research clients and sponsors.

# Preface

The RAND National Defense Research Institute assisted the National Geospatial-Intelligence Agency (NGA) in understanding how to operate in unclassified environments, including outside Sensitive Compartmented Information Facilities (SCIFs). This is the second of two reports from the project, and this report provides analysis and recommendations for intelligence agencies regarding how to conduct work outside secure government facilities by identifying policy, legal, technology, security, financial, and cultural considerations. The advantages of such changes include greater access to outside expertise, continuity of operations, and increased work-life offerings for recruitment and retention.

A companion report, *Understanding Government Telework: An Examination of Research Literature and Practices from Government Agencies*, provides a literature review and examination of telework practices at seven federal agencies.

This research was sponsored by the Human Development Directorate at NGA and conducted within the Cyber and Intelligence Policy Center of the RAND National Defense Research Institute, a federally funded research and development center sponsored by the Office of the Secretary of Defense, the Joint Staff, the Unified Combatant Commands, the Navy, the Marine Corps, the defense agencies, and the defense Intelligence Community.

For more information on the RAND Cyber and Intelligence Policy Center, see www.rand.org/nsrd/ndri/centers/intel or contact the director (contact information is provided on the webpage).

# Contents

# Tables

# Summary

This report examines how intelligence agencies can conduct more work outside classified facilities by identifying policy, legal, technology, security, financial, and cultural considerations. It further provides steps that intelligence agencies can take to address these considerations and overcome potential challenges. We reviewed studies on telework and telecommuting, examined seven federal agencies that conduct work outside government facilities, and conducted interviews inside the National Geospatial-Intelligence Agency. Intelligence agencies could benefit from conducting some unclassified functions outside Sensitive Compartmented Information Facilities (SCIFs), with each agency differing in terms of which functions would be most appropriate to move to unclassified facilities. We identified common lessons learned and recommendations for intelligence leaders to consider.

Agencies may consider transitioning employees with unclassified job functions who handle unclassified information to unclassified workplaces for several reasons. First, moving employees to unclassified environments increases opportunities to collaborate with external partners in academia, industry, national labs, and elsewhere. This could be done either by moving employees to unclassified computer networks, where they may have greater access to collaboration tools, or by allowing employees to work in geographic locations closer to industry hubs, such as Silicon Valley, or laboratories. Second, increasing the proportion of agency work that can occur on unclassified networks or in unclassified facilities aids continuity of operations during a natural disaster or act of war. Employees who have the tools, resources, and authorities to conduct their jobs remotely are less affected by events that prevent access to agency facilities. Third, if unclassified work could be conducted outside SCIFs, it is possible that agencies will need fewer employees with security clearances, thereby reducing costs associated with clearances, lag times in hiring and recruiting, and the number of insiders with access to classified information. Fourth, if agencies create systems and processes for employees to work remotely and identify unclassified functions that could occur entirely off-site, they could increase the pool of talent available for recruitment by allowing employees to work in other parts of the country or by attracting employees who may meet skill and capability requirements but not necessarily meet security clearance requirements.

Despite the potential benefits of unclassified remote work for the Intelligence Community (IC), there are also considerations that agencies will have to address when working to craft their programs. These considerations fall into six categories: policy, legal, technology, security, financial, and cultural.

Policy considerations include internal policies that agencies themselves establish that either enable or hinder remote work. If agencies establish unclassified telework programs, agencies should also establish clear, easily interpreted policies that facilitate communications between employees and supervisors; define what kind of training teleworking employees and their supervisors need; explain rules that employees should follow; and define how employees should or should not use personal and government technologies and handle sensitive information when teleworking.

Legal considerations involve laws that dictate how agencies establish their remote-work programs, such as laws on the protection of sensitive information, on the establishment of telework programs, and on the protection and handling of Controlled Unclassified Information (CUI). Our team was unable to find any specific laws governing the use of personally owned electronic devices to perform government work (except the general requirement to comply with existing security policies and regulations concerning CUI and network security), leaving each agency to decide whether employees will be allowed to log on remotely with their own computers, tablets, and smartphones and use their own printers, scanners, and other devices.

Technology considerations encompass the software and hardware that enable employees to work from unclassified facilities and remotely outside government facilities. Technology provides the practical means by which employees can perform their job functions, and we found that the technologies particularly important to IC agencies are remote log-on, online collaboration tools, and the ability to transfer files across information technology systems. Agencies should consider the following: creating remote access to desktops that will allow employees to access data necessary to their work, determining the essential but unclassified capabilities that should be moved to unclassified systems for employees' use, providing employees and managers with access to online collaboration tools, determining whether to provide government-issued computers and other devices to employees working remotely, and determining whether to allow teleworkers to use personal devices.

Security considerations involve the measures that agencies take to safeguard sensitive information. IC agencies will naturally be concerned with security because of their missions, including the protection of CUI. Agencies should therefore adopt clear, detailed, and easily understood security classification guides; implement policies on how to digitally access and handle CUI; and implement policies about how to handle and secure hard copies of CUI.

Financial considerations consist of investment costs for these changes and cost savings that may result. We found that IC agencies could reap only limited savings because of the fixed costs associated with agencies' real estate footprints and because of investment costs required to establish remote-work programs. Potential cost savings may include decreased use of utilities, fewer security clearances needed, higher employee retention rates and less money spent on hiring and training new employees, and more-efficient external collaboration processes.

Cultural considerations are the perceptions that employees and managers have of remote work, measuring the performance of remote workers, and training employees and managers on the program. We found that cultural considerations shape expectations of what remote work should be and what it can do, both for employees and for agencies as a whole. Agencies should introduce ways to make remote work acceptable to both managers and employees, while demonstrating how both groups could succeed in their careers in a culture of remote work.

How each agency's leaders address the six areas of considerations will vary from agency to agency, based on missions and current capabilities. Across all six areas, government leaders will need to decide how their agencies assess risk. The creation of an enterprise risk-management officer could help agencies develop a set of processes and tools for assessing risks, including the risk of the status quo. Agencies may find that centralizing employees in headquarters locations and relying on the classified computing system creates risks that could be mitigated by remote-work programs, even if those programs themselves introduce new risks to the agency. This report provides government leaders research and analysis from across nonintelligence agencies to inform such decisions.

## Acknowledgments

We thank Katharine Webb, adjunct policy researcher at RAND and former associate director of the RAND Intelligence Policy Center (now called the Cyber and Intelligence Policy Center), and Theodore (Ted) Clark, analytic director at CENTRA Technology, for their careful reviews of this report and the feedback they provided.

## Abbreviations

| | |
|---|---|
| BYOD | bring your own device |
| COOP | continuity of operations |
| CUI | Controlled Unclassified Information |
| DISA | Defense Information Systems Agency |
| DoD | U.S. Department of Defense |
| EO | executive order |
| FEMA | Federal Emergency Management Agency |
| FISMA | Federal Information Security Management Act of 2002 |
| GSA | General Services Administration |
| IC | Intelligence Community |
| IRS | Internal Revenue Service |
| IT | information technology |
| NASA | National Aeronautics and Space Administration |
| NGA | National-Geospatial Intelligence Agency |
| NIST | National Institute of Standards and Technology |
| NRC | Nuclear Regulatory Commission |
| OPM | U.S. Office of Personnel Management |
| PII | personally identifiable information |
| PROPIN | Proprietary Information |
| SBU | Sensitive But Unclassified |
| SCG | security classification guide |

| | |
|---|---|
| SCI | Sensitive Compartmented Information |
| SCIF | Sensitive Compartmented Information Facility |
| TEA | Telework Enhancement Act of 2010 |
| TS | Top Secret |
| VPN | virtual private network |

CHAPTER ONE

# Introduction

Intelligence agencies can learn to operate in unclassified environments, including outside of Sensitive Compartmented Information Facilities (SCIFs), by understanding certain policy, legal, technology, security, financial, and cultural considerations. To identify these considerations, we relied on a review of previous studies and research on telework and research about how federal agencies outside the Intelligence Community (IC) conduct telework. Our research on those studies and practices is available in the report *Understanding Government Telework: An Examination of Research Literature and Practices from Government Agencies.*[1]

Enabling employees to work outside of classified facilities provides several benefits to IC agencies. First, IC components have long appreciated the need for outside partnership and have frequently sought opportunities for external collaboration with experts in academia, industry, and research laboratories. In 2011, for example, the Office of the Director of National Intelligence published the *Strategic Intent for Information Sharing: 2011–2015*, which lays out the mission of improving "responsible, secure information sharing across the Intelligence Community and with external partners and customers." Its first goal toward the fulfillment of this mission is to "engage IC partners and customers (federal, military, state, local, tribal, private sector and foreign) to understand their common and unique missions needs for

[1] Cortney Weinbaum, Bonnie L. Triezenberg, Erika Meza, and David Luckey, *Understanding Government Telework: An Examination of Research Literature and Practices from Government Agencies*, Santa Monica, Calif.: RAND Corporation, RR-2023-OSD, 2018.

intelligence and information from the IC."[2] In a 2015 speech to the Professional Services Council conference, then–Director of National Intelligence James Clapper said, "Keeping up with what's on the cutting edge requires our partnership"—between the IC and the private sector. And he explained that there is a "symbiotic relationship" between government and the professional services industry.[3] However, centralizing employees inside classified headquarters buildings or isolating them from outside communities by geography or through other constraints often creates unintended hurdles to collaboration.

The Office of the Director of National Intelligence has an Office of Partner Engagement, one of whose duties is to "facilitate the sharing of appropriate and relevant intelligence information to the private sector."[4] The IC has therefore repeatedly and publicly recognized the value and necessity of external collaboration for some time. However, hurdles remain and continue to dampen the frequency of collaboration and levels of engagement. By stationing employees outside classified facilities, IC agencies could position employees closer to external partners, either virtually by moving employees and their work to unclassified computer networks or physically by enabling them to work at locations with external partners around the country without the costs associated with frequent travel. This further facilitates employees' ability to build long-term professional relationships with external partners through regular collaboration.

A second benefit to moving employees outside of classified facilities is that an increase in unclassified work could facilitate continuity of operations (COOP) during a natural disaster or an act of war. For example, many federal agencies have crafted their telework policies with COOP in mind. The Telework Enhancement Act of 2010

[2] James R. Clapper, *Strategic Intent for Information Sharing: 2011–2015*, Washington, D.C.: Office of the Director of National Intelligence, 2011.

[3] James R. Clapper, "As Prepared Remarks of DNI James Clapper Before the Professional Services Council Conference," Office of the Director of National Intelligence, October 9, 2015.

[4] Office of the Director of National Intelligence, "Partner Engagement—Who We Are," webpage, undated.

(TEA) also requires the incorporation of telework into agency COOP plans.[5] Still, using telework during a COOP event requires more than an agency policy on telework. It requires employees to understand the job functions they are allowed to conduct remotely and have access to the data and systems they need to conduct those functions, as well as to the collaboration tools to remain in communication with their own colleagues and with other offices. These elements allow employees to continue working as they normally would, even in situations where they are unable to access their regular work facilities. In a large-scale natural disaster or another kind of crisis, telework could ensure government agencies' ability to respond effectively and efficiently, mitigating the repercussions of the event on U.S. security.

Third, an increase in employees remotely conducting unclassified work could decrease the number of employees needing security clearances. This would reduce costs associated with clearance investigations, decrease hiring lag times while waiting for clearance results, and reduce insider threats by decreasing the number of people with access to classified information. If agencies could identify jobs that do not need access to classified information, and if they could provide unclassified access to the information needed for these jobs, then fewer employees would need access to classified facilities and information technology (IT) systems. Examples of jobs that could be moved off-site include analysts responsible for open-source or unclassified commercial intelligence, recruiters and other human capital officers, logisticians, procurement and contracting officers who oversee unclassified programs, and attorneys working on unclassified programs.

Fourth, if agencies were able to create systems and processes to enable employees to work remotely, and if agencies identified unclassified functions that could occur entirely off-site, they could increase the pool of talent available for recruitment to include people who might not qualify for security clearances. According to the Pew Research Center, 24.5 percent of the U.S. population (77.5 million people) was foreign

[5] Public Law 111-292, The Telework Enhancement Act of 2010, December 9, 2010.

born or born to at least one immigrant parent as of 2013,[6] meaning that a large portion of Americans may find it difficult or impossible to obtain certain levels of security clearance, regardless of their skills, education, languages, and other capabilities. In a 2014 report, the Pew Research Center found that 69 percent of adults ages 18–33 think marijuana should be legal,[7] and this population may be less inclined to apply for jobs requiring Top Secret/Sensitive Compartmented Information (TS/SCI) security clearances and agree to the restrictions—including restrictions on use of marijuana—that come with such positions. The IC currently has an employment model where TS/SCI security clearances are generally required for employment. This is a one-size-fits-all approach rather than specific to the job, because the vast majority of information necessary for IC work is available only on classified computing systems. If the IC could attract talent that does not meet TS/SCI security clearance requirements for many human resource management functions, open-source collection and analysis, and other similar jobs, it could increase the pool of potential candidates and reduce security investigation and onboarding costs. Further, intelligence agencies could offer telework as a retention option to current employees. Especially among millennials, a high proportion of workers prefer flexible workspaces outside the traditional office.[8]

This report includes six topics for IC leaders when considering these programs: policy, legal, technology, security, financial, and cultural considerations. Chapter Two discusses policy considerations: the internal policies agencies establish that either enable or hinder remote work. Chapter Three discusses legal considerations: the relevant laws, statutes, and executive orders that agencies should consider when designing their policies and programs. Chapter Four discusses tech-

---

[6] Pew Research Center, "Overview," in *Second-Generation Americans: A Portrait of the Adult Children of Immigrants*, Washington, D.C., February 7, 2013.

[7] Pew Research Center, "Generations and Issues," in *Millennials in Adulthood: Detached from Institutions, Networked with Friends*, Washington, D.C., March 7, 2014.

[8] Cortney Weinbaum, Richard Girven, and Jenny Oberholtzer, *The Millennial Generation: Implications for the Intelligence and Policy Communities*, Santa Monica, Calif.: RAND Corporation, RR-1306-OSD, 2016.

nology considerations: the various tools, including software and hardware, that agencies could consider to facilitate remote work. Chapter Five discusses security considerations: measures that agencies can take to safeguard sensitive information. Chapter Six discusses financial considerations: possible cost savings or financial investments that may result from such programs. Chapter Seven discusses cultural considerations: factors that affect a workforce's attitudes toward telework, including perceptions, how to measure performance, training, and the use of telework as a recruitment and retention tool. We found multiple examples of topics that overlap across these six topics, and we discuss those overlaps when they arise. Each chapter includes our findings and recommendations for the topic, and we provide final conclusions in Chapter Eight.

## Methodology

We examined how various federal agencies enable employees to work with sensitive information outside government facilities. Across government, telework was the most common approach for remote work, and telework is the approach with the most public data available. Other approaches for remote work exist but are implemented differently at each agency that uses them, making research difficult using existing data. This report adds to RAND's research in the report *Understanding Government Telework: An Examination of Research Literature and Practices from Government Agencies*.[9] In that report, RAND staff collaborated with graduate students from Syracuse University's Maxwell School of Citizenship and Public Affairs to review literature on remote work and the telework programs at seven federal agencies.

We then conducted 44 interviews with National-Geospatial Intelligence Agency (NGA) officials to understand the hurdles that employees and managers face when trying to implement remote work. We interviewed officials across NGA's mission offices and support functions to examine how policy, legal, security, technology, and financial

---

[9] Weinbaum et al., 2018.

considerations affect remote work. During our analysis, we identified additional lessons learned that did not fit in any of these topics; we therefore created a sixth area, called cultural considerations, which we define in Chapter Seven.

The rules set by the TEA require that teleworkers work from an "approved worksite," and this requirement might not be desired by agencies that want their employees to have the ability to work from anywhere at any time. In this report, we use the term *remote work* to describe work that may occur anywhere outside a government facility because this term does not assume that employees are working from an approved worksite. Telework is used far less frequently within the IC than in nonintelligence agencies,[10] and this report provides options and recommendations for intelligence agencies that want to increase their use of telework or other remote-work practices.

---

[10] In 2016, 22 percent of all federal employees teleworked, while 51 percent of those eligible to telework did. Similar figures do not exist either for the IC as a whole or for individual IC agencies. However, a U.S. Office of Personnel Management (OPM) telework report noted that 179 employees of the Central Intelligence Agency teleworked in FY 2016. For more information, see U.S. Office of Personnel Management (OPM), *Status of Telework in the Federal Government: Report to Congress; Fiscal Year 2016*, Washington, D.C., November 2017, pp. 5, 83.

CHAPTER TWO

# Policy Considerations

Policies shape the guidelines for how agencies operate, including by delineating roles, responsibilities, and processes and by defining rules to explain how programs and laws should be implemented. We found that when agencies lack policies that clearly explain how employees can effectively work remotely, managers and employees hesitate to utilize remote work as an option, or they may implement it in ways that are not appropriate. For intelligence agencies specifically, clear policies provide an opportunity for agency leaders to describe missions and functions well suited for working remotely, appropriate security practices when working remotely, appropriate levels of engagement across employees and with managers, appropriate collaboration practices, and acceptable uses of technology. When policies are unclear or incomplete, employees and managers may interpret them in conflicting ways, which can result in the underutilization of remote-work practices or of remote workers, or it can result in inappropriate use of remote work.

We found specific areas where clear policy guidance could facilitate effective remote work at intelligence agencies:

1. **Telework policy**: Agencies desiring to use telework are required to have a telework policy that, according to OPM, "should be written in such a way that it can be clearly understood and easily used."[1] A telework policy should, at a minimum, cover the topics on which the TEA empowers OPM to provide guid-

[1] OPM, *Guide to Telework in the Federal Government*, Washington, D.C., April 2011, p. 10.

ance and consultation, such as pay and leave, agency closure, performance management, official worksite, recruitment and retention, and accommodations for employees with disabilities.[2] OPM's *Guide to Telework in the Federal Government* provides further details about what items should be included in a telework policy, grouped into three broad categories: definitions of eligibility and location, requirements and expectations for training employees and managers, and requirements for record keeping and reporting.[3] The importance of such a policy is that it establishes the option to telework while also regulating how it is to be carried out. As the *Guide to Telework in the Federal Government* states, "Telework is primarily an arrangement established to facilitate the accomplishment of work."[4]

2. **Use of personal devices**: Agencies should have a policy on the use of government-issued or personal devices to perform work. OPM guidance states that telework policies should "reference agency information technology (IT) and cybersecurity guidelines."[5] Agencies therefore have leeway but also the responsibility to craft policies that conform to their missions and needs. Chapter Five discusses legal aspects of this topic.
3. **Handling of sensitive information**: Because of the nature of their missions, IC components need detailed, but also easily comprehensible, policies about how employees should handle different types of information, including Controlled Unclassified Information (CUI), personally identifiable information (PII), and Proprietary Information (PROPIN). Chapters Three and Five discuss this topic in greater detail from the legal and security perspectives.
4. **Core work hours**: OPM's *Guide to Telework in the Federal Government* notes that specifying the days of the week and hours

---

[2] Pub. L. 111-292, 2010, p. 3.

[3] OPM, 2011, pp. 10–13.

[4] OPM, 2011, p. 11.

[5] OPM, 2011, p. 11.

to be worked during telework days falls under telework agreements. Although the TEA does not require agencies to include specific days and hours in their telework agreements, OPM recommends that they do so.[6] Agencies could establish guidelines for core work hours, such as by defining how many days within a pay period may be used for telework. That will condition expectations and provide employees and managers a common reference point when they begin to negotiate telework agreements in greater detail.

5. **Time and attendance management**: Agencies should establish policies about how to record time and attendance when employees are off-site. For many federal agencies, such a policy is not necessary because time and attendance is remotely accessible from anywhere, yet IC agencies may need a policy if their time and attendance system is on a classified computing network that employees cannot access remotely.

The next section provides examples and lessons learned from several government agencies on how to implement policies to facilitate remote work.

## Lessons from Across Government

The TEA offers some guidance regarding what should be included in agencies' telework policies. For example, it empowers OPM to "provide policy and policy guidance for telework in the areas of pay and leave, agency closure, performance management, official worksite, recruitment and retention, and accommodations for employees with disabilities."[7] The TEA also stipulates that the head of each executive agency shall "determine the eligibility for all employees of the agency to participate in telework" and ensure that "an interactive telework training program is provided to (A) employees eligible to participate in the

---

[6] OPM, 2011, p. 17.

[7] Pub. L. 111-292, 2010, p. 3.

telework program of the agency; and (B) all managers of teleworkers."[8] Agencies' telework polices should, therefore, cover at least these topics that the TEA emphasizes. In addition to these minimum requirements, we found commonalities across agencies that have robust telework programs, including policies that facilitate better, more-frequent dialogue between employees and managers; mandate extra training that is specifically tailored to positions or situations; and outline more opportunities for evaluating teleworking employees.

The General Services Administration (GSA) is an example of an organization whose policies facilitate dialogue between employees and managers. The *GSA Mobility and Telework Policy* says, "Good performance management practices, including appropriate formal and informal feedback, are essential for all employees to work effectively."[9] GSA hosts central station meetings to present changes and solicit feedback on the telework policy. Such meetings are open to all GSA employees with "an interest in participating and/or providing feedback to the policy process."[10]

GSA, the Nuclear Regulatory Commission (NRC), NGA, and the Federal Emergency Management Agency (FEMA) also have policies mandating training tailored to positions and situations. The first three agencies mandate separate training for employees and supervisors, primarily using programs developed by OPM.[11] FEMA's telework manual states, "Employees and supervisors and all co-workers of participating employees should attend the FEMA Telework training program."[12] FEMA has further designed a COOP telework exercise called Determined Sentry that can be used to "determine current capabilities to operate in a telework or socially-distanced environment

[8] Pub. L. 111-292, 2010, pp. 1–3.

[9] GSA, *GSA Mobility and Telework Policy*, GSA Order, Update HCO 6040.1, Washington, D.C., undated, p. 7.

[10] GSA, *The GSA Telework Program Management Office Recipe Book*, Washington, D.C., November 2011.

[11] Telework.gov, OPM, "Training & Resources," webpage, undated-d; GSA, "Resources for Managing Teleworkers," webpage, last reviewed August 13, 2017a.

[12] FEMA, *FEMA Manual 3000.3*, Washington, D.C., July 2000.

and to determine what needs to be done to enhance [FEMA's] current capabilities and to better prepare for a pandemic influenza or continuity event."[13]

FEMA has also instituted a number of opportunities for evaluating teleworkers. According to its telework manual, regular teleworkers are subject to at least six evaluations each year of their participation in the telework program. Moreover, "after the first year of participation, and each subsequent year, a brief recertification process will be necessary to continue in the Telework Program for the following year."[14] This regular rate of evaluation not only facilitates communication but also helps maintain a healthy level of expectations among employees and managers. Managers know how their teleworking employees are performing, while the employees know that they are being credited for their work.

## Recommendations for Intelligence Agencies

IC agencies should implement policies that do the following:

- Facilitate communication between employees and supervisors in terms of deciding which positions are eligible for telework, when persons in these positions can perform telework, and what kinds of tasks can be performed off-site.
- Define what kind of training employees and their supervisors need, especially so that the former will be able to continue performing competently even when working remotely.
- Explain the rules that employees must follow, including how time and attendance should be recorded, how productivity will be evaluated, and any required core work hours or required work processes.

---

[13] FEMA, *Department and Agency: Continuity Telework Exercise; Exercise Plan (EXPLAN) Template*, Washington, D.C., May 2013.

[14] FEMA, 2000, p. 17.

- Describe how employees should or should not use personal and government technologies and how they should handle sensitive information when teleworking. Clear policies will help agencies adhere to the Federal Records Act of 1950 and prevent data spills of classified information on unclassified computing systems.[15]

[15] Public Law 81-754, Federal Records Act of 1950, September 5, 1950.

CHAPTER THREE

# Legal Considerations

This chapter identifies federal statutes and executive orders that should be considered by intelligence agencies when assessing how employees could work remotely. These statutes and other directives will inform how agencies craft and refine policies and how they implement various aspects of a remote-work program. In this chapter, we assume that IC agencies have already established secure IT networks according to National Institute of Standards and Technology (NIST) standards, and we include only statutes and executive orders (EOs) that apply to employees rather than to contractors. The analysis in this chapter is provided to inform agency leaders, and it is not intended to replace the legal guidance provided by agency general counsels.

We found three regulations to be most relevant to intelligence agencies implementing remote-work programs and a fourth topic where we searched but did *not* find regulations:

1. **Federal Information Security Management Act of 2002 (FISMA)** mandates information security controls over information resources that support federal operations and assets.
2. **The TEA** defines *telework* and mandates that agencies establish specific policies, processes, and training for employees and managers.
3. **EO 13556** creates the CUI Program to replace previous unclassified control markings and establishes standard procedures for handling CUI across agencies.
4. **Use of personally owned devices** for unclassified government work is not forbidden by law, and agencies have flexibility in

creating their own policies and processes for use of employee-owned devices, such as using personal smartphones and laptops to remotely log into government systems.

## Lessons from Across Government

### Federal Information Security Management Act of 2002

FISMA established a "framework for ensuring the effectiveness of information security controls over information resources that support Federal operations and assets."[1] FISMA requires federal agencies to ensure the security of federal government data and further requires the head of each agency to conduct annual information security reviews. IC agency leaders must ensure that all personnel, including those who are permitted to work remotely, comply with the IC agency's information security program.

In 2014, Congress amended FISMA to accomplish additional federal government information security–related objectives, several of which are important for IC agencies that permit employees to work remotely.[2] The amendments to FISMA gave oversight authority to the director of the Office of Management and Budget (OMB) for "agency information security policies and practices," and it gave the Secretary of the U.S. Department of Homeland Security the authority "to administer the implementation of such policies and practices for information systems." In addition, the amendment provided the following:

- "OMB's information security authorities are delegated to the Director of National Intelligence (DNI) for certain systems operated by an element of the intelligence community."[3]

---

[1] Public Law 107-347, Title III, Federal Information Security Management Act of 2002, December 17, 2002.

[2] Public Law 113-283, Federal Information Security Modernization Act of 2014, Section 3553, December 18, 2014.

[3] Congressional Research Service, summary of Federal Information Security Modernization Act of 2014, S. 2521, Washington, D.C., 2014.

- Agency heads must ensure that "all personnel are held accountable for complying with the agency-wide information security program."[4]

An agency's information security programs may need to be revised to address specific information security measures that apply to employees who are working remotely. NIST has published guidance for agencies to comply with the requirements of FISMA.[5]

### Telework Enhancement Act of 2010

The TEA defines *telework* as "a work flexibility arrangement under which an employee performs the duties and responsibilities of such employee's position, and other authorized activities, from an approved worksite other than the location from which the employee would otherwise work."[6]

The act directs each head of an agency to establish a policy for telework, determine which employees are eligible to telework, and notify all employees of the agency of their eligibility to telework. The act requires that a contract for telework be established between an agency manager and an employee who is authorized to telework that outlines the arrangements for teleworking. The act further requires that the head of each executive agency establish an "interactive training program" for employees who are eligible to telework and their managers.[7]

The TEA also establishes security guidelines to "ensure the adequacy of information and security protections for information and information systems used while teleworking."[8] It directs that "[e]ach executive agency shall incorporate telework into the continuity of oper-

---

[4] Congressional Research Service, 2014.

[5] Joint Task Force Transformation Initiative, *Security and Privacy Controls for Federal Information Systems and Organizations*, Washington, D.C.: National Institute of Standards and Technology, NIST SP 800-53, April 2013.

[6] Pub. L. 111-292, 2010, Section 6501(3).

[7] Pub. L. 111-292, 2010, Section 6502 (a)(1) (A), (B), and (C).

[8] Pub. L. 111-292, 2010, Section 6504 (c).

ations plan of that agency,"[9] and the "head of each executive agency shall designate an employee of the agency as the Telework Managing Officer."[10] Finally, the act outlines various reports on the participation of employees in teleworking and other data that must be submitted to various federal agencies.[11] This legislation requires agency heads to implement the act without any stated exemptions for IC agencies.

Similarly, the Federal Management Regulation, in Subpart F, provides regulations that address telework policies for executive-branch agencies.[12] These regulations were established prior to the TEA, pursuant to the Department of Transportation and Related Agencies Appropriations Act, 2001.[13] These regulations address, for example, the federal facility telework policy that executive agencies must follow;[14] the steps that agencies must take to implement those laws and policies;[15] and how agencies can obtain guidance, assistance, and oversight regarding workplace arrangements from GSA.[16] The provisions of Section 102-74.585 are congruent with the TEA, in that they require that "each Executive agency . . . establish a policy under which eligible employees of the agency may participate in telecommuting to the maximum extent possible without diminished employee performance."[17]

---

[9] Pub. L. 111-292, 2010, Section 6504 (d).

[10] Pub. L. 111-292, 2010, Section 6505.

[11] Pub. L. 111-292, 2010, Section 6506.

[12] Code of Federal Regulations, Title 41, Chapter 102, Federal Management Regulation, Part 102-74.585–102-74.600, 2011.

[13] Public Law No. 106-346, Department of Transportation and Related Agencies Appropriations Act, 2001, 114 Stat. 1356, October 31, 2001.

[14] Code of Federal Regulations, 2011, Part 102-74.585 (2011).

[15] Code of Federal Regulations, 2011, Part 102-74.590.

[16] Code of Federal Regulations, 2011, Part 102-74.595.

[17] Code of Federal Regulations, 2011, Part 102–74.585.

**Executive Order 13556: Controlled Unclassified Information**

The final rule concerning CUI became effective on November 14, 2016.[18] The new regulation was derived from EO 13556, *Controlled Unclassified Information*, issued by President Barack Obama. The EO was issued to "establish an open and uniform program for managing [unclassified] information that requires safeguarding or dissemination controls."[19] The final rule states that, prior to the issuance of the EO, "more than 100 different markings for such information existed across the executive branch." It noted that "this ad hoc agency-specific approach created inefficiency and confusion, led to a patchwork system that failed to adequately safeguard information requiring protection, and unnecessarily restricted information-sharing."

The new regulation established the CUI Program "to standardize the way the executive branch handles information that requires safeguarding or dissemination controls (excluding information that is classified under Executive Order 13526, Classified National Security Information, 75 FR 707 (December 29, 2009), or the Atomic Energy Act of 1954 (42 U.S.C. 2011, as amended))."[20] The regulation states that "the Federal Government's Executive Agent (EA) for Controlled Unclassified Information (CUI)" is the "National Archives and Records Administration (NARA), through its Information Security Oversight Office (ISOO)."[21] It further states:

> All unclassified information throughout the executive branch that requires any safeguarding of dissemination control is CUI. Law, regulation (to include this part), or government-wide policy must require or permit such controls. Agencies therefore may not

[18] Code of Federal Regulations, Title 32, Part 2002, Controlled Unclassified Information, November 14, 2016; EO 13556 of November 4, 2010, Controlled Unclassified Information, *Federal Register*, Vol. 75, No. 216, November 9, 2010.

[19] Code of Federal Regulations, 2016.

[20] Code of Federal Regulations, 2016. For the sources cited in the quote, see EO 13526 Classified National Security Information, *Federal Register*, Vol. 75, No. 2, January 5, 2010, p. 707; and United States Code, Title 42, Section 2011, Congressional Declaration of Policy, October 24, 1992.

[21] Code of Federal Regulations, 2016.

> implement safeguarding or dissemination controls for any unclassified information other than those controls consistent with the CUI Program.[22]

The CUI Program sets safeguarding standards for CUI that must be followed by employees who work remotely and that may be difficult for agencies to achieve consistently. For example, the EO requires the following:

> 1. Establish controlled environments in which to protect CUI from unauthorized access or disclosure and make use of those controlled environments;
>
> 2. Reasonably ensure that unauthorized individuals cannot access or observe CUI, or overhear conversations discussing CUI;
>
> 3. Keep CUI under the authorized holder's direct control or protect it with at least one physical barrier, and reasonably ensure that the authorized holder or the physical barrier protects the CUI from unauthorized accessor observation when outside a controlled environment; and
>
> 4. Protect the confidentiality of CUI that agencies or authorized holders process, store, or transmit on Federal information systems in accordance with the applicable security requirements and controls established in FIPS PUB [Federal Information Processing Standards Publication] 199, FIPS PUB 200, and NIST SP [Special Publication] 800-53 (incorporated by reference, see Section 2002.2) and paragraph (g) of this section.[23]

Certain items in this list are more difficult to achieve for employees who live with other adults and may require the employee to have locked storage for any printed materials. It also may be difficult for agency heads to monitor and enforce compliance with all of the CUI safeguards by employees working off-site. One approach to complying

[22] Code of Federal Regulations, 2016, Section 2002.1(c).

[23] Code of Federal Regulations, 2016, Section 2002.14(c).

with this requirement could be inspection of employees' workspaces, which is what the NRC does by sending inspectors to teleworkers' home offices.

The new regulation sets out requirements for marking, reproducing, and destroying CUI. A training program for all agency employees can address the safeguards and protocols for the CUI Program. Particular attention should be given to the challenges of complying with the CUI safeguards and program requirements faced by employees who work remotely.

**Personally Owned Devices**

The RAND team searched for statutes and regulations that govern the use of personally owned devices at government agencies, but we were unable to identify any specific requirements for "bring your own device" (BYOD) or personally owned devices other than compliance with existing agency security policies and the statutes and regulations discussed previously. Employees working remotely may use a variety of devices for work—such as computers, smartphones, and tablets to read email, communicate, and create and share documents—and each agency establishes its own policies for personally owned devices.

A recent NIST publication, *User's Guide to Telework and Bring Your Own Device (BYOD) Security*, explains that "telework devices can be divided into two categories: personal computers (desktops, laptops) and mobile devices (e.g., smartphones, tablets)." Sometimes, the agency owns the telework device, but "telework devices controlled by the user are also known as *bring your own device (BYOD)*."[24] The NIST publication addresses the risks posed by the use of telework devices controlled by the user and offers practical guidance to users and agencies about how to minimize those risks. Specifically, the NIST publication explains the critical importance of properly securing BYOD devices, which are very popular with teleworking employees:

[24] Murugiah Souppaya and Karen Scarfone, *User's Guide to Telework and Bring Your Own Device (BYOD) Security*, Washington, D.C.: National Institute of Standards and Technology, NIST Special Publication 800-114, Revision 1, July 2016, p. vi.

> Many organizations limit the types of BYOD devices that can be used and which resources they can use, such as permitting BYOD laptops to access a limited set of resources and permitting all other BYOD devices to access webmail only. This allows organizations to limit the risk they incur from BYOD devices. When a telework device uses remote access, it is essentially a logical extension of the organization's own network. Therefore, if the telework device is not secured properly, it poses additional risk to not only the information that the teleworker accesses but also the organization's other systems and networks. For example, a telework device infected with a worm could spread the worm through remote access to the organization's internal computers. Therefore, telework devices should be secured properly and have their security maintained regularly.[25]

IC agencies that consider whether employees should be permitted to use their own devices should make these evaluations in the context of the requirements of the laws previously discussed.

## Recommendations for Intelligence Agencies

This chapter has outlined some of the federal legislation and legal considerations that IC agencies should include when evaluating how to permit employees to work remotely:

- Protect information security using policies and practices that meet FISMA requirements.
- When using telework as the process for remote work, implement the requirements of the TEA.
- Implement policies and practices that adhere to the CUI Program and the NIST BYOD guide.[26]

---

[25] Souppaya and Scarfone, 2016, p. vi.

[26] Souppaya and Scarfone, 2016.

CHAPTER FOUR

# Technology Considerations

Technology is an enabler for remote work in that it provides the practical means by which employees can work off-site. Technology allows employees to access the data necessary to do their jobs, including working with emails and important files, while also providing collaboration tools that allow employees to remain in contact with colleagues and supervisors regardless of their work locations. For intelligence agencies concerned about security and insider threats, technology can also be used to monitor for malicious user activity and detect external threats. Technology itself may entail certain risks, but with appropriate security measures in place, it can serve as an indispensable tool for conducting remote work.

We found overlap between how agencies address technology needs and how they approach security considerations when specific technologies are not available to employees because of security decisions made by agencies. For example, certain file-sharing capabilities, video conferencing, and data-transfer capabilities are all technically feasible and commercially available but may not be available to intelligence personnel because agencies have deemed these capabilities too risky to use. Agencies can review their approaches toward collaborative technologies that enable employees to work off-site by weighing the risk of using these technologies against the costs of not using them and by considering how improvements to agency policies and training could allow certain technologies to be used safely.

Our research across other federal agencies found that having the right kinds of technology greatly facilitates remote work for both employees and managers. Such technology might include hardware (such as laptops, remote log-on access tokens, and phones) and software (such as virtual private networks [VPNs], web applications, and collaboration tools). Technologies that stand out as particularly important to intelligence agencies include the following:

1. **Remote log-on**: The ability of employees to access their unclassified emails, applications, and files from outside government facilities. This may include VPN secure computing environments to provide safeguards to prevent improper access of CUI and monitor for malicious activity.
2. **Collaboration tools**: The ability to collaborate and communicate remotely may include instant message and chat room capabilities, phone-call forwarding, and the ability to share files across teams. In cases where IC agencies block access to commercial tools because of security restrictions, they should provide suitable alternatives on their own unclassified networks.
3. **File transfer across IT systems**: Intelligence personnel who rely on classified computing systems often store their unclassified documents on classified systems. This becomes a hurdle to remote work because employees are unable to remotely access these files. This hurdle can be overcome by instituting responsive transfer processes—subject to multilevel approval to ensure continued security—to efficiently move files between classified and unclassified systems. Such processes might include assigning individuals or groups of individuals the responsibility of transferring files for other employees.

The following section provides examples of how several government agencies have used various technologies to facilitate remote work.

## Lessons from Across Government

We found that the technology considerations that contribute to the success of telework programs are those that

- allow employees access to their agencies' unclassified intranet, email, and other unclassified information to do their jobs.

- provide access to collaboration tools for all employees irrespective of location so they can talk and share information with each other
- provide policies and guidelines about the use of unclassified technologies
- determine which devices agencies will issue to employees, such as government laptops or common access card (CAC) readers, and any associated costs.

We found that several agencies have set up remote desktops that teleworking employees can access from anywhere. The NRC uses an application called Citrix, which "allows employees remote computer access to NRC systems and applications."[1] Other agencies, such as NGA and the Coast Guard, also have their own remote desktops that teleworking employees can use.[2]

The National Aeronautics and Space Administration (NASA) uses a range of commercially available software to help teleworkers stay in contact with their coworkers, including Skype (for voice and video calls), Adobe Connect (for sharing audio, video, chat, references, and screens), Google Docs (for collaboratively writing, editing, and storing documents in real time), Asana (for monitoring employee productivity), and Dropbox (for hosting and accessing files on a cloud).[3] GSA uses Online Meeting, which "allows participants from varying locations [to] all view the same presentation on their computer screen, while editing and revising documents," and Voice over Internet Protocol (VoIP) to make phone calls over the internet.[4]

To mitigate security risks, FEMA and NASA have multiple encryptions and impose such limitations as requiring multiple permissions before granting teleworkers access to technologies. At FEMA, telework employees wishing to access their emails must have a Terminal

---

[1] NRC, Office of the Inspector General, *Audit Report: Audit of NRC's Telework Program*, Washington, D.C., OIG-10-A-13, June 9, 2010, p. 5.

[2] U.S. Coast Guard, Office of Civilian Human Resources, "Telework Frequently Asked Questions," February 3, 2014, p. 7.

[3] Ali Llewellyn, "Teleworking," *Open NASA*, November 15, 2011.

[4] GSA, "Tools for Effective Telework," webpage, last reviewed August 13, 2017c.

Access Controller Access Control System (TACACS), which requires permission to obtain. These employees must also obtain permission from relevant organizations within FEMA before they can access CUI and records subject to the Privacy Act of 1974.[5]

NASA forbids the use of personal computers at NASA headquarters or through the headquarters VPN, and all NASA data must be accessed through either a government-approved or a government-issued computer. Additional requirements include (1) VPN software installed for the computer, (2) internet access from a third-party internet service provider, and (3) a personal identification verification smartcard (NASA Badge) and PIN for logging into the computer and VPN.[6]

Agencies can take steps to encourage remote work by providing various kinds of technology-related incentives. According to its telework manual, FEMA will take into account the individual needs of participants and may provide telephone services, communications, computers, agency-owned equipment, and other supplies.[7] The Defense Information Systems Agency (DISA), budget permitting, may also reimburse teleworkers for 50 percent of "their expense incurred for the installation and monthly fee for commercially provided high-speed internet service."[8]

The use of personal devices poses risks and concerns to many agencies. The Department of the Navy's telework policy states, "Use of personally-owned equipment, such as a personal computer (PC), for telework is authorized as a last resort only when GFE [government-furnished equipment] is not provided or available."[9] The U.S. Department of the Interior also "prefers that employees use government owned equipment for teleworking. However, the use of personally

---

[5] FEMA, 2000, pp. 15, 19.

[6] NASA, "Remote Access Services," webpage, Information Technology and Communications Division, undated-a.

[7] FEMA, 2000, pp. 15–16.

[8] DISA Instruction 220-15-57, *Administration: Telework*, Washington, D.C.: Defense Information Systems Agency, December 6, 2012, p. 3.

[9] Secretary of the Navy Instruction 12271.1, *Department of the Navy Telework Policy*, Washington, D.C.: U.S. Department of the Navy, Office of the Secretary, May 12, 2015, p. 3.

[owned] equipment (POE) is permissible when certain conditions are met." These conditions include having "government-approved encryption security for sensitive information" and keeping "all government information separate from your personal information," such as through a separate drive or folder.[10]

As teleworking becomes more commonplace, the topic of personal devices will become more acute as agencies are faced with financial decisions associated with providing employees with laptops or other devices. Certain agencies have managed to arrive at compromise solutions that allow employees to use personal devices, thereby saving agencies money, but that still maintain security standards. At NASA, for instance, "[p]ersonal mobile devices that access NASA data . . . are required to have the Agency baseline Active Sync policies applied. With these security features and policies in place, personal mobile devices present no greater risk than what is accepted for a government-issued mobile device."[11] Other organizations have come up with more technologically innovative solutions. For example, the Department of Homeland Security's cyber research division has funded a "new Bluetooth-enabled mobile thin client," which allows a user to "click into an app on her phone or tablet and connect to a totally separate set of apps and information stored in a data center or computer cloud, not leaving any information on the phone itself."[12] Such technology could be applied to other types of mobile devices as well. There are ways to overcome the problems that personal devices present.

---

[10] U.S. Department of the Interior, "FS12-07: Use of Personally Owned Equipment for Teleworking," fact sheet, Washington, D.C., January 2012, p. 1.

[11] NASA, "Telework Preparedness FAQ," Information Technology and Communications Division, undated-b.

[12] Joseph Marks, "DHS Showcases Bluetooth Tech at RSA," *Nextgov*, February 17, 2017.

## Recommendations for Intelligence Agencies

IC agencies should consider technologies and packages of technology solutions that can contribute to remote-work programs, including the following:

- Create remote access to allow employees to access email and other programs and data necessary to their work.
- Consider which essential and unclassified capabilities should move from classified to unclassified systems so that employees will be able to access these capabilities irrespective of work location.
- Make available to employees collaborative tools they can use from any location, including instant messaging, group-chat capabilities, file sharing, video conferencing, workflow tracking and project management, and calendar sharing.
- Decide whether to provide government-issued laptops and other hardware to employees who work remotely or whether to rely on personally owned devices. If agencies allow teleworkers to use personal devices, then they will need to make sure that the proper security measures are in place.

CHAPTER FIVE

# Security Considerations

Security is critical for IC agencies, given the nature of their work, and has historically been a roadblock to the development of robust remote-work programs and, in general, to working at the unclassified level. There is a belief within some areas of the IC that agencies *must* do work within classified environments or that working within classified environments is the only available option. There are, in fact, many areas where intelligence agencies do unclassified work, and that work could occur outside of classified facilities. We found examples of government agencies mitigating security risks through policies, practices, and tools that could be implemented in conjunction with a risk assessment. Topics that intelligence agencies may consider include the following:

1. **Security classification guides (SCGs)**: These guides tell employees what information is classified, at various levels, and explains handling restrictions. Yet SCGs are often vague, confusing, or not all inclusive of the information employees handle. Agencies should consolidate their SCGs so that they conform to new government-wide standards mandated by EO 13556 and the rule for enforcing it, enacted by the Information Security Oversight Office of the National Archives and Records Administration.[1] Agencies should further clarify their SCGs to incorporate delineations for when information crosses from one security classification level to another, including informing employees

[1] EO 13556, 2010; Ronald D. Lee, Dana E. Koffman, and Tom McSorley, "Inside the New Controlled Info Rule for Contractors," *Law 360*, October 11, 2016.

how to determine whether classification by compilation is a risk (when information that independently is unclassified becomes classified when combined). Such guides should clearly explain which information requires CUI special handling.

2. **Electronic data handling policies**: These policies include guidance on how employees may or may not access CUI via email (government email versus personal email), remote desktops, and other methods and under which circumstances employees must use encryption and digital signatures.
3. **Physical handling policies**: These policies include guidance on how employees may or may not remove printed CUI from facilities, print hard copies outside of government facilities, secure information when off-site, and destroy copies that are no longer needed. Such policies may require the use of locked drawers, shredders, or other equipment.

This is not an exhaustive list. It is merely intended to begin a dialogue about how intelligence agencies could approach designing safe remote-work programs. Our research and analysis were developed based on the assumption that intelligence agencies already have existing security protocols in place, including computer-based monitoring for malicious activity and insider threats and continuous monitoring procedures for employees with access to sensitive information.

## Lessons from Across Government

We found that security policies, practices, and training are important to remote-work programs, including (1) detailed, yet simple and easily interpreted, SCGs; (2) policies and training for how employees should handle CUI on IT systems; and (3) policies on how to handle, secure, and destroy hard copies of CUI.

The NRC, whose employees handle nuclear information, provides detailed, easily interpreted security classification guidelines. It has three that are particularly relevant to its telework program. First, the NRC has the manual *Sensitive Unclassified Information Security*

*Program*, which covers such topics as how different types of documents should be marked, how they are supposed to be physically secured, and how Sensitive But Unclassified (SBU) information should be transmitted over telecommunications systems.[2] The NRC also has the *Glossary of Security Terms*, which provides definitions of the different kinds of CUI.[3] The NRC has separate guidance on safeguards information, a type of information that is unclassified but important to the type of work that the NRC does.[4]

The U.S. Department of Defense (DoD) Education Activity has its own telework policy but also refers to the overall DoD telework policy when it comes to handling CUI on IT systems. DoD's policy for handling PII states that teleworkers may access PII only on encrypted government systems with two-factor authentications and that PII can be emailed only between government accounts and "must be encrypted and digitally signed."[5]

The NRC's directive on sensitive unclassified information security also deals with the physical security of such information. Although the directive originally focused on the securing of sensitive information at NRC facilities, it has since been applied to doing the same at telework locations. For example, the directive requires that information must be stored in repositories that would provide adequate physical protection, including but not limited to steel filing cabinets with steel locking bars, three-position changeable combinations, padlocks, and other features; security filing cabinets that have been certified to certain standards; and bank safe deposit boxes.[6] Inspectors visit the remote worksites of

---

[2] NRC, *NRC Sensitive Unclassified Information Security Program*, Washington, D.C., Directive 12.6, DT-99-33, June 2, 1998.

[3] NRC, *Glossary of Security Terms*, Management Directive 12.0, Washington, D.C., DT-14-17, July 1, 2014b.

[4] NRC, *NRC Safeguards Information Security Program*, Washington, D.C., Management Directive 12.7, June 27, DT-14-16, 2014a.

[5] Department of Defense Instruction 1035.01, *Telework Policy*, Washington, D.C.: U.S. Department of Defense, April 4, 2012, p. 16.

[6] NRC, 1998.

teleworkers, usually their homes, to check that employees have cabinets with locks and combinations.[7]

Although vague security policies could lead to misinterpretation, overly detailed guides may cause their own problems. According to Senator Dianne Feinstein, current member and former chair of the Senate Select Committee on Intelligence, as of 2016, "the Information Security Oversight Office found more than 3,000 classification guides in existence across the U.S. government, including more than 2,000 in the Defense Department."[8] This overabundance of guidance creates a tremendous amount of (sometimes conflicting) information that can confuse and overwhelm employees and managers. In 2010, the White House noted the problem of nonuniform classification standards across government. It observed that agencies across government employed ad hoc, agency-specific classification, creating an "inefficient, confusing patchwork [that] has resulted in inconsistent marking and safeguarding of documents" and other problems. For this reason, President Obama issued EO 13556, which provides that "CUI categories and subcategories shall serve as exclusive designations for identifying unclassified information throughout the executive branch that requires safeguarding or dissemination controls," replacing For Official Use Only (FOUO), SBU, PII, and other previous classifications.[9]

## Recommendations for Intelligence Agencies

IC agencies could adopt a number of security measures and policies that will contribute to secure remote work:

- Write clear SCGs that explain what information is CUI and that describe other handling caveats in accordance with the new government-wide standards mandated by EO 13556. IC agencies

---

[7] Weinbaum et al., 2018.

[8] Dianne Feinstein, "How to Rethink What's 'Top Secret' for the Internet Age," *Washington Post*, December 16, 2016.

[9] EO 13556, 2010, p. 1.

should also remove conflicts between different security guides to prevent the mishandling of information because of inconsistent rules and interpretations. This will allow employees to identify and differentiate what is and is not CUI and the appropriate channels for handling the information.

- Implement policies about how to digitally access and handle CUI. This should cover such topics as how to access this information and how to transfer it, including when to use encryption and digital signatures.
- Implement policies about how to physically handle and secure CUI and consider providing any necessary equipment, such as locked drawers.

CHAPTER SIX

# Financial Considerations

The financial implications of creating remote-work programs will differ across various government agencies, and each agency should consider both the potential cost savings and any new financial investments required to establish a program. Our research indicates that, within the IC, agencies may experience more financial costs from creating remote-work programs than they do financial savings. This chapter describes financial considerations for IC agencies, specifically the following:

1. **Investment costs**: The cost to implement the necessary features of a successful remote-work environment may include new technology investments—possibly purchasing government laptops or other equipment for employees to use when working remotely. Agencies could also expect there to be personnel time to draft new policies, create and implement training, and create feedback mechanisms to monitor the program and employee compliance, as well as other costs associated with a successful policy shift.
2. **Financial savings:** The most significant cost saving that federal agencies report from remote work is the ability to consolidate facilities to reduce the square footage of office space needed and lower utility costs. These savings may not be possible for intelligence agencies whose workforces are centralized in headquarters facilities. Other savings are possible through reduced employee attrition and the need for fewer security clearances, if agencies adopt remote-work programs that allow some employees to work entirely without security clearances. Some agencies

measure increased productivity as a saving, and these benefits are available to agencies whose workforces continue to work through weather events.

3. **Employee costs and savings**: Employees themselves may incur costs and savings from remote-work programs, and these create both hurdles and incentives for employees to participate in these programs. The largest cost that employees will experience is setting up an effective workspace, which may include new computing equipment and peripherals if the agency does not provide these. Employees may be required to upgrade their home internet connectivity or invest in new office furniture, monitors, filing cabinets, phones, printers, and so on that they otherwise would not purchase. Yet employees will experience savings in reduced time and costs to commute, and such time savings may benefit employees who are the primary caretaker for a family member.

## Lessons from Across Government

Data on the initial cost of implementing telework policies for federal agencies are not publicly available and likely vary by agency, depending on time of implementation, size of the initial program, and the security requirements of the agency. Agencies may implement the technology infrastructure needed to work remotely before actually deciding to implement a program, leading to investment costs not directly associated with the remote-work program. The most-common costs we found associated with federal remote-work programs were laptops, and some agencies also describe providing employees with printers, scanners, and locked drawers. We did not find cost estimates for how much agencies spend on these types of equipment.[1]

Cost savings associated with telework are generally from an agency's need for less office space. Agencies that consolidate facilities or reallocate office space after implementing telework programs may

---

[1] Weinbaum et al., 2018.

report savings in real estate costs and utilities. Table 6.1 provides examples of financial savings from real estate consolidations following the implementation of agencies' remote-work programs.

We are unable to determine whether intelligence agencies may reap similar financial benefits, because many intelligence agencies have consolidated their Washington-area employees into headquarters facilities, limiting their options to reduce their real estate footprints. In cases where agencies are examining new facility options—such as NGA's new facility in St. Louis, Missouri—lessons from other agencies that have consolidated workspaces or used shared workspaces (called *hoteling*) could lead to financial savings. At GSA, where the majority of employees telework (88 percent of employees in 2015)[2] and do not need

**Table 6.1**
**Examples of Cost Savings from Telework**

| Agency | Savings | Description of Savings |
|---|---|---|
| Chemical Safety and Hazard Investigation Board | $500,000 per year | The agency reduced real estate costs by $500,000 a year by downsizing its real estate footprint. |
| FEMA | $9.6 million | FEMA reduced office leases by consolidating most of its headquarters employees through expanding participation in telework and implementing desk-sharing. |
| Department of the Treasury | $5.4 million | Three of Treasury's bureaus have realized substantial cost savings, of $5,367,015, associated with building closures or consolidations where telework was a vehicle to support the project. |
| GSA | $30.4 million | GSA reported $24.6 million in annual rent savings and $6 million in annual administrative cost savings. |
| United States Patent and Trade Office | $38.2 million | The United States Patent and Trade Office avoided spending $38.2 million to secure additional office space. |

SOURCES: OPM, 2017; FEMA data came from Brittany Ballenstedt, "FEMA Ramps Up Telework, Mobility," *Nextgov*, May 10, 2013.

[2] OPM, 2017.

offices, the administration was able to create an efficient and adaptive open-plan office space, leading to the significant annual cost savings.

Some agencies have reported impacts on retention because of telework, which provides cost savings through decreased employee turnover rates (and associated costs for recruitment, onboarding, and training). When federal agencies set goals for their telework programs, 35 percent of agencies set goals for recruitment and 35 percent set goals for retention.[3] Those decisions may be worthwhile, because during the 2016 Federal Employee Viewpoint Survey, employees who teleworked described higher rates of employee engagement than employees who did not.[4] An analysis of 11 research papers revealed that the cost of replacing a worker in the private sector is 21 percent of the employee's salary.[5] If teleworking increases retention rates, the increase in retention saves agencies money. The *NASA Desk Guide on Telework Programs* lists some of the advantages of teleworking as "recruit and retain high-quality talent," "improve employee morale and a better balance of work and personal lives," and "reduce commuting related stress and costs."[6] The Internal Revenue Service (IRS) handbook for human resources lists some benefits as "enhanced recruitment and retention especially in critical occupations and positions," "increased employee productivity and job satisfaction," and "improved work-life balance."[7]

The Departments of Agriculture and Defense have used telework as a recruitment tool, and the Department of Treasury, the International Boundary and Water Commission, the United States Patent and Trade Office, the National Transportation Safety Board, and the Farm

---

[3] OPM, 2017, p. 8.

[4] OPM, *The Keys to Unlocking Engagement: An Analysis of the Conditions That Drive Employee Engagement*, Washington, D.C., 2016, Table 5.

[5] Heather Boushey and Sarah Jane Glynn, "There Are Significant Business Costs to Replacing Employees," Center for American Progress, November 16, 2012.

[6] NASA, *NASA Desk Guide on Telework Programs*, Washington, D.C.: Office of Human Capital Management, NSREF-3000-0012, April 2010, p. 7.

[7] IRS, "Part 6: Human Resources Management, Chapter 800.0 Employee Benefits, Section 2: Telework (Flexiplace) Program," in *Internal Revenue Manuals*, Washington, D.C., last reviewed September 10, 2017.

Credit System Insurance Corporation have used telework to retain top talent.[8] DISA also used telework as a way to retain employees when it consolidated facilities and relocated to Fort Meade because of Base Realignment and Closure (BRAC).[9] Many federal agencies that use telework to meet retention goals describe specific groups of employees within their goals, such as employees with personal and family needs or employees eligible for retirement.[10]

In a global survey of 2,912 adults between the ages of 18 and 50 working full-time in white-collar roles, 45 percent of respondents agreed with the statement, "I want the freedom to work and play from anywhere at anytime with no restrictions," and 27 percent agreed with the statement, "I would be willing to accept a pay cut in return for greater work flexibility."[11] Therefore, remote work and telework may be useful tools to attract and retain this portion of the workforce that wants flexibility and is willing to make financial concessions to receive it.

Telework has been theorized to affect employee productivity through reduced use of sick leave, either because the flu and colds cannot spread as efficiently through the workforce or because employees can work remotely when they feel sick. We have been unable to find data either supporting or refuting these hypotheses, but we did find that *medical telework* is an approved use of telework at several government agencies.[12]

Another impact on productivity is the ability of employees to work remotely when they otherwise might be unable to work. In February 2010, a blizzard in the Washington metro area closed the federal

---

[8] Telework.gov, OPM, "Common Myths," webpage, undated-a; Telework.gov, OPM, "Recruitment & Retention," webpage, undated-b.

[9] Doug Beizer, "DISA Will Use Telework to Retain Displaced Employees," *FCW*, September 4, 2009.

[10] OPM, 2017, p. 21.

[11] No respondents worked in IT, research and development, engineering, or human resources; respondents did not work for a company in the education, market research, tech services or consulting, or nonprofit industries; all respondents had a managerial or administrative role; and all respondents worked for an organization that employs at least 100 people worldwide. Cisco, *2014 Connected World Technology Final Report*, San Jose, Calif., 2014.

[12] OPM, 2017, p. 21.

government for four full days, led to an early dismissal the day the storm arrived, and resulted in two days of delayed arrival or unscheduled leave after the storm ended.[13] NASA estimated that its Work from Anywhere program saved $30 million for each day the government was closed because of snowstorms that year,[14] demonstrating a measurable example of increased productivity. In January 2016, during Winter Storm Jonas, GSA's telework program allowed 3,600 out of 3,800 GSA employees to continue to work in the Washington metro area while 20 inches of snow fell in the region.[15]

If the financial costs of security clearances are included in this analysis, then remote work becomes a significant driver for cost savings. First, if remote-work programs improve retention, even slightly, then the government saves the costs of investigating and clearing the new would-be replacement employees. Second, when more positions in intelligence agencies become eligible for remote work and agencies create the infrastructure to enable remote work, some jobs may be conducted entirely off-site. Such a change could lead to groups of employees no longer needing access to classified data and computing systems, reducing the needs for their Top Secret security clearances and reducing the risk of insider threats associated with clearing those employees. Third, remote work could reduce the need for overhiring (hiring more employees than are necessary to backfill vacant positions or replace employees on temporary leave) by moving specific job functions off-site.

The costs and savings that employees themselves incur are important for agency leaders to understand because they may affect the willingness of employees to participate in a remote-work program and whether the agency will achieve its desired goals. Americans spend an

---

13 OPM, "Snow & Dismissal Procedures: Status Archives," webpage, undated.

14 NASA, "Work from Anywhere: Factsheet," undated-c.

15 David Shive, "Federal Workforce on Duty During Winter Storm Jonas," *GSA Blog*, January 28, 2016; Jason Samenow, "How Much Snow Fell from Snowzilla in the D.C. Area, in Detail," *Washington Post*, January 24, 2016.

average of 200 hours and $2,600 per year on their daily commutes,[16] so teleworking might save employees time and money on commuting. We found that many federal agencies issue laptops and other equipment (as necessary) to employees for the dual purpose of removing financial barriers to remote work and to provide the level of security desired on computing devices and storage containers (file cabinets, safes, etc.). Decisions regarding what equipment to issue to employees versus what equipment to require the employees themselves to provide could result in additional costs for the workforce.

## Recommendations for Intelligence Agencies

IC agencies could consider the financial impacts of a remote-work program through the following actions:

- Build cost estimates for the initial investments required for a new remote-work program and the long-term costs to sustain the program. Initial costs may include developing training programs or installing new IT infrastructure, and long-term costs may include the maintenance of government-issued laptops.
- Set goals for cost savings that include estimates for the number of employees who would work remotely and how the agency could adapt to this new structure. If the agency anticipates reducing office space, plans should include how this would be accomplished and what savings could be anticipated.
- Consider costs that employees would incur or savings that employees may reap, and communicate these expectations to employees choosing to participate in the program.

---

[16] Kathryn Vassel, "We Spend $2,600 a Year Commuting to Work," *CNN Money*, June 17, 2015.

CHAPTER SEVEN

# Cultural Considerations

IC agencies considering or expanding programs for remote work or telework should consider factors that do not easily fall within any of the considerations discussed in previous chapters. Many of the employees and managers we interviewed described these things as *culture*, and in this chapter we discuss what that term meant to them. We did not initially intend to study the cultural impacts of remote work, and therefore this chapter is not a complete assessment of all topics that may be described as cultural considerations. However, throughout our research on the previous five topics, we found factors that shape expectations of what remote work should be and what it can do, both for employees and for employers. These topics did not fit in any of the previous sections:

1. **Perceptions**: The perceptions of employees and managers toward telework and other forms of remote work affect whether they see these practices as effective or not. If employees believe that remote work will have a negative impact on their careers, or if managers believe that remote work will lead to a decline in the quality of employees' work, then either factor could become a hurdle to participation. By including remote-work goals in performance reviews and promotion criteria, agency leaders can demonstrate how these programs advance careers and benefit the agency's mission.
2. **Measuring performance**: Agency leaders and managers need the ability to evaluate the productivity of remote workers, similar to how they would evaluate on-site workers. The mentality of

"if I can't see you, you're not working for me" can be overcome by management practices (such as assigning tasks based on deliverables rather than time), frequent communication (such as using collaboration tools), and other forms of technology (such as software that monitors productive and idle computing time). Each agency would need to choose the set of solutions that fits best with its missions, functions, and culture.

3. **Training**: Agencies with telework programs are required to provide telework training to managers and employees. Additional training on results-oriented or deliverable-based management practices, the management of teams and collaborating across geographic distances, and topics relevant to the mission and functions of the agency's remote workers could improve how remote-work programs are used.

## Lessons from Across Government

In our research, we found that perceptions often influenced the degree of participation in telework programs at federal agencies. A workforce that values regular, in-person contact among employees and managers can be hesitant or even outright hostile toward remote work. Agency workforces may believe that teleworkers are "out of sight, out of mind," which OPM lists as one of the most common myths about telework.[1] (Several of these myths are listed in the box on the next page.) Perceptions of remote work may include managers believing that remote workers are either doing their jobs poorly or not at all and employees fearing that managers who hold these beliefs may penalize them in performance reviews. If employees believe that they will be passed over for promotions or miss opportunities to build workplace connections if they choose to telework, they may not want to pursue telework opportunities.

Such perceptions have to be countered from the top, with buy-in from senior leaders and then with midlevel managers. Indeed, accord-

[1] Telework.gov, OPM, undated-a.

**Myths About Federal Telework**

"I will not know my employees are working at home."

"Teleworkers are not available when you need them."

"Telework is not for everyone, so it is not fair."

"Everyone will want to telework."

"Teleworkers cause more work for supervisors."

"Teleworkers cause more work for coworkers."

"Teleworkers are out of sight, out of mind."

SOURCE: Telework.gov, OPM, "Common Myths," webpage, undated-a.

ing to OPM, "Research in the work/life field bears out that supervisors, managers and senior executives who model the use of workplace flexibilities, such as telework, serve as key drivers in effecting positive cultural change in that organization."[2]

The ability to measure performance affects whether negative perceptions of remote work can be countered. Managers may be concerned about effectively managing workers remotely, measuring or evaluating productivity, and being held accountable to the performance reviews they give. Generally, government guidelines have focused on results rather than process when it comes to remote work. These guidelines stress the importance of providing employees with deliverable-based assignments. Sometimes, agencies may also complement this approach with certain kinds of technology.

OPM stresses, "In a telework environment, managing by results and not by physical presence becomes even more critical."[3] By this logic, managers could focus on issuing deliverable-based or task-based assignments, thereby mitigating the need to closely monitor teleworkers' time. This is the approach that the IRS takes. According to the

---

[2] Telework.gov, OPM, undated-a.

[3] Telework.gov, OPM, "Results-Oriented Management," webpage, undated-c.

guidelines for its telework program (Flexipace), "Remote work management, including telework, requires managers to be more diligent in managing for results and focusing on outcomes, business results and deliverables." The five steps for managing teleworkers are planning work and setting expectations, monitoring performance, developing employee skills, appraising performance, and recognizing employees for their accomplishments.[4] The Defense Logistics Agency says, "Clearly established schedules and shared expectations regarding work to be performed, deliverables, and timelines can help minimize the possibility of employees working unauthorized overtime hours or taking undocumented leave."[5] GSA states, "Manage by results, not by physical presence. . . . Establish a clear definition of objectives and performance indicators, and ensure close monitoring of those indicators along with ongoing training for teleworking employees."[6]

Some agencies have used technology to keep track of the activities and performance of remote workers. GSA has dashboards that track teleworkers' performance and provide metrics that allow cross-regional comparisons, although the GSA website does not specify exactly how these dashboards do the former.[7] NASA uses a commercially available management application called Asana, which allows "teams to keep track of who is doing what and where they are investing [their] time and efforts."[8]

We found that training is a way to familiarize employees and managers with telework. Training helps both parties better understand their responsibilities in such an arrangement and will also shape their expectations—of what they should expect both from telework and from each other. A diverse range of agencies, including NGA, DoD Education Activity, and the National Institutes of Health, currently

---

4 IRS, 2017.

5 Defense Logistics Agency Instruction 7212, *Defense Logistics Agency (DLA) Telework Program*, Washington, D.C.: Defense Logistics Agency, February 26, 2013.

6 GSA, "Resources for Managing Teleworkers," webpage, last reviewed August 13, 2017a.

7 GSA, 2017d.

8 Llewellyn, 2011.

make use of the Telework 101 modules for employees and managers that OPM provides on its telework website.[9] Agencies may also provide training that is agency-specific, just as the National Institutes of Health and GSA do.[10] GSA's training is publicly available and consists of 15 short narrated videos that discuss such topics as work sites or the changing culture at GSA. Several of these videos include interviews with GSA teleworkers who discuss their experiences in the program.

## Recommendations for Intelligence Agencies

IC agencies should take the following steps to shape attitudes toward remote work in a more positive way:

- Communicate changes to managers and employees, including how the changes fit within the agency mission, what changes will occur, who is eligible for off-site work, and what roles each part of the organization has in these changes.
- Provide measures for the program's performance and tools for managers to determine employee performance and to effectively manage off-site employees. These mechanisms may involve new policies, manager training, or technology tools.
- Demonstrate that employees will remain upwardly mobile along their career paths by celebrating promotions of remote workers, performance awards received by remote workers, and other career accomplishments achieved by remote workers.

---

[9] National Institutes of Health, "Telework Training," webpage, Office of Human Resources, undated; Department of Defense Education Activity, "Telework," webpage, undated.

[10] GSA, *Telework Training*, Washington, D.C., last reviewed August 13, 2017b.

CHAPTER EIGHT

# Conclusions

Moving specific job functions or entire jobs out of the SCIF could enable intelligence agencies to geographically distribute their workforces to improve collaboration with organizations not located near agency facilities, enhance preparations for disaster events, and reduce risks from insider threats. But such changes to the business model of an agency require leadership involvement and several concrete steps. The agency's mission needs and the desired effects of changing the business model (distributed workforce, cost reduction, flexibility in hiring, etc.) should be part of the decision as to whether employees who handle unclassified information should work off-site part of the time (in a telework arrangement that requires them to come to the office weekly) or all of the time (in a remote-work arrangement where they would not have access to classified information or need to maintain a security clearance). The advantages and disadvantages to the agency and the employee should be weighed against the financial costs and nonfinancial costs of such a change.

We found that agencies wanting to plan such changes need to consider such factors as how much new technology infrastructure the agency would need to create and whether the agency could fully fund the program in the first year or would need to take an incremental approach. These types of factors would influence the amount of time an agency would need to implement a remote-work program. Agencies whose leaders are interested in this change would first need to decide who should be responsible, because many of the considerations described in this report would cross organizational boundaries within the agency.

Across the agencies we examined, leaders had to balance costs and cost savings, mission benefits, and risks to agency operations. The decision criteria that agencies used to make these assessments were not readily available in the open-source literature. For some agencies, an enterprise risk management officer may assist in developing processes and tools for leaders across organizational units to discuss and assess risks, including the risks of their current operating procedures. The types of risks that intelligence leaders may need to address while deciding whether to move employees off-site include the following:

- Is the agency missing access to specific critical skills or other talents by limiting recruitment to candidates willing to work on-site, and how should this be weighed against the costs of creating a remote-work program?
- How should agencies weigh the costs and logistical burden of providing employees with government-issued laptops against the risks of allowing employees to connect to the network from their own devices?
- In preparing for a COOP event, how should agencies balance the risk-mitigation benefits of dispersing employees across geographic locations and across the power grid and other critical infrastructure against the benefits created by the hardened systems that agencies may have available on-site?
- Would segmenting the workforce into those who have access to classified information and those who do not create a two-tiered workforce?

These questions must be addressed in a systematic manner by leaders at each agency. This report provides the initial set of considerations agency leaders should weigh during their analyses, and all of the recommendations proposed throughout this report are consolidated in Table 8.1.

IC agencies looking to establish or expand remote-work programs will need to address each of the areas discussed in this report—policy, legal, technology, security, financial, and cultural—while tailoring each area to their own specific missions and current capabilities.

**Table 8.1**
**Recommendations**

| Topic | Recommendations |
| --- | --- |
| Policy | • Facilitate communications between employees and supervisors, particularly in terms of defining who is eligible for telework, when they can perform such telework, and what kinds of tasks can be performed off-site.<br>• Define what kind of training employees and their supervisors need, especially so that the former will be able to continue performing competently even when working remotely.<br>• Explain rules that employees must follow, including how time and attendance should be recorded, how productivity will be evaluated, and any required core work hours or required work processes.<br>• Describe how employees should or should not use personal and government technologies and how they should handle sensitive information when teleworking. Clear policies will help agencies adhere to the Federal Records Act and prevent data spills of classified information on unclassified computing systems. |
| Legal | • Protect information security using policies and practices that meet FISMA requirements.<br>• When using telework as the process for remote work, implement the requirements of the TEA.<br>• Implement policies and practices that adhere to the CUI Program and NIST BYOD guide. |
| Technology | • Create remote access to allow employees to access email and other programs and data necessary to their work.<br>• Consider which essential and unclassified capabilities should move from classified to unclassified systems so that employees will be able to access them irrespective of work location.<br>• Make available to employees collaborative tools they can use from any location, including instant messaging, group chat capabilities, file sharing, video conferencing, workflow tracking and project management, and calendar sharing.<br>• Decide whether to provide government-issued laptops and other hardware to employees who work remotely or whether to rely on personally owned devices. If agencies allow teleworkers to use personal devices, then they will need to make sure that the proper security measures are in place. |

**Table 8.1—Continued**

| Topic | Recommendations |
|---|---|
| Security | • Write clear SCGs that explain what information is CUI and that describe other handling caveats in accordance with the new government-wide standards mandated by EO 13556 (2010). Also remove conflicts between different security guides to prevent mishandling of information from inconsistent rules and interpretations. This will allow employees to identify and differentiate what is and is not CUI and the appropriate handling channels.<br>• Implement policies on how to digitally access and handle CUI. This should cover such topics as how to access this information and how to transfer it, including when to use encryption and digital signatures.<br>• Implement policies on how to physically handle and secure CUI, and consider providing any necessary equipment, such as locked drawers. |
| Financial | • Build cost estimates for the initial investments required for a new remote-work program and the long-term costs to sustain the program. Initial costs may include developing training programs or installing new IT infrastructure, and long-term costs may include the maintenance of government-issued laptops.<br>• Set goals for cost savings that include estimates for the number of employees who would work remotely and how the agency could adapt to this new structure. If the agency anticipates reducing office space, plans should include how this would be accomplished and what savings could be anticipated.<br>• Consider costs that employees would incur or savings that employees may reap, and communicate these expectations to employees choosing to participate in the program. |
| Cultural | • Communicate changes to managers and employees, including how the change fits within the agency mission, what changes will occur, who is eligible for off-site work, and what roles each part of the organization has in this change.<br>• Provide measures for the program's performance and tools for managers to determine employee performance and to effectively manage employees off-site. These mechanisms may include new policies, manager training, or technology tools.<br>• Demonstrate that employees will remain upwardly mobile along their career paths by celebrating promotions of remote workers, performance awards received by remote workers, and other career accomplishments achieved by remote workers. |

# References

Ballenstedt, Brittany, "FEMA Ramps Up Telework, Mobility," *Nextgov*, May 10, 2013. As of March 13, 2017:
http://www.nextgov.com/cio-briefing/wired-workplace/2013/05/fema-ramps-telework-mobility/63111/

Beizer, Doug, "DISA Will Use Telework to Retain Displaced Employees," *FCW*, September 4, 2009. As of November 26, 2017:
https://fcw.com/articles/2009/09/04/disa-telework.aspx

Boushey, Heather, and Sarah Jane Glynn, "There Are Significant Business Costs to Replacing Employees," Center for American Progress, November 16, 2012. As of November 26, 2017:
https://www.americanprogress.org/issues/economy/reports/2012/11/16/44464/there-are-significant-business-costs-to-replacing-employees/

Cisco, *2014 Connected World Technology Final Report*, San Jose, Calif., 2014.

Clapper, James R., *Strategic Intent for Information Sharing: 2011–2015*, Washington, D.C.: Office of the Director of National Intelligence, 2011. As of November 26, 2017:
https://www.dni.gov/files/documents/Strategic%20Intent%20for%20Information%20Sharing.pdf

———, "As Prepared Remarks of DNI James Clapper Before the Professional Services Council Conference," Office of the Director of National Intelligence, October 9, 2015. As of November 26, 2017:
https://www.dni.gov/index.php/newsroom/speeches-interviews/speeches-interviews-2015/item/1264-as-prepared-remarks-of-dni-james-clapper-before-the-professional-services-council-conference

Code of Federal Regulations, Title 41, Chapter 102, Federal Management Regulation, Part 102-74.585–102-74.600, 2011.

———, Title 32, Part 2002, Controlled Unclassified Information, November 14, 2016.

Congressional Research Service, summary of Federal Information Security Modernization Act of 2014, S. 2521, Washington, D.C., 2014. As of January 8, 2018:
https://www.congress.gov/bill/113th-congress/senate-bill/2521

Defense Information Systems Agency Instruction 220-15-57, *Administration: Telework*, Washington, D.C.: Defense Information Systems Agency, December 6, 2012.

Defense Logistics Agency Instruction 7212, *Defense Logistics Agency (DLA) Telework Program*, Washington, D.C.: Defense Logistics Agency, February 26, 2013. As of November 26, 2017:
http://www.dla.mil/Portals/104/Documents/J5StrategicPlansPolicy/PublicIssuances/i7212.pdf

Department of Defense Education Activity, "Telework," webpage, undated. As of February 17, 2017:
https://content.dodea.edu/teach_learn/operations/onboarding_ed_directorate/policies_procedures_telework.html

Department of Defense Instruction 1035.01, *Telework Policy*, Washington, D.C.: U.S. Department of Defense, April 4, 2012.

DISA—*See* Defense Information System Agency.

EO—*See* Executive Order.

Executive Order 13556 of November 4, 2010, Controlled Unclassified Information, *Federal Register*, Vol. 75, No. 216, November 9, 2010. As of November 26, 2017:
https://www.gpo.gov/fdsys/pkg/FR-2010-11-09/pdf/2010-28360.pdf

Federal Emergency Management Agency, *FEMA Manual 3000.3*, Washington, D.C., July 2000. As of November 26, 2017:
https://www.fema.gov/pdf/library/3000_3.pdf

———, *Department and Agency: Continuity Telework Exercise; Exercise Plan (EXPLAN) Template*, Washington, D.C., May 2013. As of November 26, 2017:
https://www.fema.gov/media-library/assets/documents/86274

Feinstein, Dianne, "How to Rethink What's 'Top Secret' for the Internet Age," *Washington Post*, December 16, 2016. As of November 26, 2017:
https://www.washingtonpost.com/opinions/how-to-rethink-whats-top-secret-for-the-internet-age/2016/12/16/57835e0e-bccd-11e6-94ac-3d324840106c_story.html?utm_term=.fb72c01d5943

FEMA—*See* Federal Emergency Management Agency.

General Services Administration, *GSA Mobility and Telework Policy*, GSA Order, Update HCO 6040.1, Washington, D.C., undated. As of January 25, 2017:
https://www.gsa.gov/portal/mediaId/122806/fileName/GSAteleworkpolicy.action

———, *The GSA Telework Program Management Office Recipe Book*, Washington, D.C., November 2011.

———, "Resources for Managing Teleworkers," webpage, last reviewed August 13, 2017a. As of January 25, 2017:
https://www.gsa.gov/portal/category/102107

———, *Telework Training*, Washington, D.C., last reviewed August 13, 2017b. As of November 26, 2017:
https://www.gsa.gov/portal/category/102551

———, "Tools for Effective Telework," webpage, last reviewed August 13, 2017c. As of November 26, 2017:
https://www.gsa.gov/portal/content/288381

———, "Telework," webpage, last reviewed October 16, 2017d. As of November 26, 2017:
https://www.gsa.gov/portal/category/26435

GSA—*See* General Services Administration.

Internal Revenue Service, "Part 6: Human Resources Management, Chapter 800.0 Employee Benefits, Section 2: Telework (Flexiplace) Program, in *Internal Revenue Manuals*, Washington, D.C., last reviewed September 10, 2017. As of November 26, 2017:
https://www.irs.gov/irm/part6/irm_06-800-002.html

IRS—*See* Internal Revenue Service.

Joint Task Force Transformation Initiative, *Security and Privacy Controls for Federal Information Systems and Organizations*, Washington, D.C.: National Institute of Standards and Technology, NIST SP 800-53, April 2013.

Lee, Ronald D., Dana E. Koffman, and Tom McSorley, "Inside the New Controlled Info Rule for Contractors," *Law 360*, October 11, 2016. As of November 26, 2017:
https://www.law360.com/articles/849107/inside-the-new-controlled-info-rule-for-contractors

Llewellyn, Ali, "Teleworking," *Open NASA*, November 15, 2011. As of March 21, 2018:
http://www.opennasa.org/teleworking.html

Marks, Joseph, "DHS Showcases Bluetooth Tech at RSA," *Nextgov*, February 17, 2017. As of November 26, 2017:
http://www.nextgov.com/mobile/2017/02/dhs-showcases-bluetooth-tech-rsa/135554/

NASA—*See* National Aeronautics and Space Administration.

National Aeronautics and Space Administration, "Remote Access Services," webpage, Information Technology and Communications Division, undated-a. As of January 26, 2017:
https://www.hq.nasa.gov/office/itcd/remote_access.html

———, "Telework Preparedness FAQ," Information Technology and Communications Division, undated-b. As of February 28, 2017:
https://www.hq.nasa.gov/office/hqhr/docs/FAQ%20Remote%20connectivity.pdf

———, "Work from Anywhere: Factsheet," undated-c. As of February 20, 2017:
http://www.hq.nasa.gov/office/hqhr/docs/Factsheet.pdf

———, *NASA Desk Guide on Telework Programs*, Washington, D.C.: Office of Human Capital Management, NSREF-3000-0012, April 2010. As of November 26, 2017:
https://searchpub.nssc.nasa.gov/servlet/sm.web.Fetch/Telework_Desk_Guide.pdf?rhid=1000&did=778067&type=released

National Institutes of Health, "Telework Training," webpage, Office of Human Resources, undated. As of February 13, 2017:
https://hr.od.nih.gov/workingatnih/telework/training.htm

NRC—*See* U.S. Nuclear Regulatory Commission.

Office of the Director of National Intelligence, "Partner Engagement—Who We Are," webpage, undated. As of November 29, 2017:
https://www.dni.gov/index.php?option=com_content&view=article&id=402&Itemid=813

OPM—*See* U.S. Office of Personnel Management.

Pew Research Center, "Overview," in *Second-Generation Americans: A Portrait of the Adult Children of Immigrants*, Washington, D.C., February 7, 2013. As of November 26, 2017:
http://www.pewsocialtrends.org/2013/02/07/second-generation-americans/

———, "Generations and Issues," in *Millennials in Adulthood: Detached from Institutions, Networked with Friends*, Washington, D.C., March 7, 2014. As of November 26, 2017:
http://www.pewsocialtrends.org/2014/03/07/millennials-in-adulthood/

Public Law 81-754, Federal Records Act of 1950, September 5, 1950.

Public Law 106-346, Department of Transportation and Related Agencies Appropriations Act, 2001, October 23, 2000.

Public Law 107-347, Title III, Federal Information Security Management Act of 2002, December 17, 2002.

Public Law 111-292, The Telework Enhancement Act of 2010, December 9, 2010.

Public Law 113-283, Federal Information Security Modernization Act of 2014, December 18, 2014.

Samenow, Jason, "How Much Snow Fell from Snowzilla in the D.C. Area, in Detail," *Washington Post*, January 24, 2016.

Secretary of the Navy Instruction 12271.1, *Department of the Navy Telework Policy*, Washington, D.C.: U.S. Department of the Navy, Office of the Secretary, May 12, 2015. As of November 26, 2017:
http://govdocs.rutgers.edu/mil/navy/12271.1.pdf

Shive, David, "Federal Workforce on Duty During Winter Storm Jonas," *GSA Blog*, January 28, 2016. As of March 13, 2017:
https://gsablogs.gsa.gov/gsablog/2016/01/28/federal-workforce-on-duty-during-winter-storm-jonas/

Souppaya, Murugiah, and Karen Scarfone, *User's Guide to Telework and Bring Your Own Device (BYOD) Security*, Washington, D.C.: National Institute of Standards and Technology, NIST Special Publication 800-114, Revision 1, July 2016.

Telework.gov, U.S. Office of Personnel Management, "Common Myths," webpage, undated-a. As of February 27, 2017:
https://www.telework.gov/federal-community/telework-managers/common-myths/

———, "Recruitment & Retention," webpage, undated-b. As of February 17, 2017:
https://www.telework.gov/guidance-legislation/telework-guidance/recruitment-retention/

———, "Results-Oriented Management," webpage, undated-c. As of February 27, 2017:
https://www.telework.gov/federal-community/telework-managers/results-oriented-management/

———, "Training & Resources," webpage, undated-d. As of on January 25, 2017:
https://www.telework.gov/training-resources/

U.S. Coast Guard, Office of Civilian Human Resources, "Telework Frequently Asked Questions," February 3, 2014. As of March 6, 2018:
http://www.dcms.uscg.mil/Portals/10/CG-1/cg121/docs/Benefits/Telework_FAQ.pdf?ver=2017-03-23-142747-920

U.S. Department of the Interior, "FS12-07: Use of Personally Owned Equipment for Teleworking," fact sheet, Washington, D.C., January 2012. As of November 26, 2017:
https://www.doi.gov/sites/doi.gov/files/migrated/telework/upload/Personally-Owned-Equipment.pdf

U.S. Nuclear Regulatory Commission, *NRC Sensitive Unclassified Information Security Program*, Washington, D.C., Directive 12.6, DT-99-33, June 2, 1998. As of November 26, 2017:
https://www.nrc.gov/docs/ML0417/ML041700603.pdf

———, *NRC Safeguards Information Security Program*, Washington, D.C., Management Directive 12.7, DT-14-16, June 27, 2014a. As of November 26, 2017: https://www.nrc.gov/docs/ML1414/ML14142A166.pdf

———, *Glossary of Security Terms*, Management Directive 12.0, Washington, D.C., DT-14-17, July 1, 2014b. As of November 26, 2017: https://www.nrc.gov/docs/ML1013/ML101380472.pdf

U.S. Nuclear Regulatory Commission, Office of the Inspector General, *Audit Report: Audit of NRC's Telework Program*, Washington, D.C., OIG-10-A-13, June 9, 2010. As of January 29, 2018: https://www.nrc.gov/docs/ML1016/ML101600394.pdf

U.S. Office of Personnel Management, "Snow & Dismissal Procedures: Status Archives," webpage, undated. As of March 13, 2017: https://www.opm.gov/policy-data-oversight/snow-dismissal-procedures/status-archives/#year2010

———, *Guide to Telework in the Federal Government*, Washington, D.C., April 2011. As of November 26, 2017: https://www.telework.gov/guidance-legislation/telework-guidance/telework-guide/guide-to-telework-in-the-federal-government.pdf

———, *The Keys to Unlocking Engagement: An Analysis of the Conditions That Drive Employee Engagement*, Washington, D.C., 2016. As of March 13, 2017: https://www.fedview.opm.gov/2016FILES/Keys_Unlocking_Engagement.pdf

———, *Status of Telework in the Federal Government: Report to Congress; Fiscal Year 2016*, Washington, D.C., November 2017. As of March 6, 2018: https://www.telework.gov/reports-studies/reports-to-congress/2017-report-to-congress.pdf

Vassel, Kathryn, "We Spend $2,600 a Year Commuting to Work," *CNN Money*, June 17, 2015. As of November 26, 2017: http://money.cnn.com/2015/06/17/pf/work-commute-time-and-money/

Weinbaum, Cortney, Richard Girven, and Jenny Oberholtzer, *The Millennial Generation: Implications for the Intelligence and Policy Communities*, Santa Monica, Calif.: RAND Corporation, RR-1306-OSD, 2016. As of November 26, 2017: http://www.rand.org/pubs/research_reports/RR1306.html

Weinbaum, Cortney, Bonnie L. Triezenberg, Erika Meza, and David Luckey, *Understanding Government Telework: An Examination of Research Literature and Practices from Government Agencies*, Santa Monica, Calif.: RAND Corporation, RR-2023-OSD, 2018. As of April 30, 2018: http://www.rand.org/pubs/research_reports/RR2023.html